THIS LOGBOOK BELONGS TO:

NAME:

_____
_____
_____

ADDRESS & PHONE:

_____
_____
_____

EMAIL:

_____
_____
_____

TRUCKER LOG

COMPANY: _____ PHONE: _____

TRUCK NO. _____ DRIVER: _____

TRAILER NO. _____ DRIVER: _____

DATE LEFT: _____ DATE RETURNED: _____

TRIP FROM: **TRIP TO:**

DATE	STATE	ODOMETER BEGINNING MILES	ODOMETER ENDING MILES	HIGHWAYS USED EACH STATE	STATE MILES

FUEL PURCHASE RECORD

DATE	NUMBER	GALLONS	COST	NAME	CITY	STATE	CASH/CHARGE CREDIT CARD

🚛 TRUCKER LOG

COMPANY: _____   PHONE: _____

TRUCK NO. _____   DRIVER: _____

TRAILER NO. _____   DRIVER: _____

DATE LEFT: _____   DATE RETURNED: _____

TRIP FROM: **TRIP TO:**

DATE	STATE	ODOMETER BEGINNING MILES	ODOMETER ENDING MILES	HIGHWAYS USED EACH STATE	STATE MILES

FUEL PURCHASE RECORD

DATE	NUMBER	GALLONS	COST	NAME	CITY	STATE	CASH/CHARGE CREDIT CARD

◄▬▬▬ TRUCKER LOG

COMPANY: _____	PHONE: _____
TRUCK NO. _____	DRIVER: _____
TRAILER NO. _____	DRIVER: _____
DATE LEFT: _____	DATE RETURNED: _____

TRIP FROM: **TRIP TO:**

DATE	STATE	ODOMETER BEGINNING MILES	ODOMETER ENDING MILES	HIGHWAYS USED EACH STATE	STATE MILES

FUEL PURCHASE RECORD

DATE	NUMBER	GALLONS	COST	NAME	CITY	STATE	CASH/CHARGE CREDIT CARD

🚛 TRUCKER LOG

COMPANY: _____   PHONE: _____

TRUCK NO. _____   DRIVER: _____

TRAILER NO. _____   DRIVER: _____

DATE LEFT: _____   DATE RETURNED: _____

TRIP FROM: TRIP TO:

DATE	STATE	ODOMETER BEGINNING MILES	ODOMETER ENDING MILES	HIGHWAYS USED EACH STATE	STATE MILES

FUEL PURCHASE RECORD

DATE	NUMBER	GALLONS	COST	NAME	CITY	STATE	CASH/CHARGE CREDIT CARD

TRUCKER LOG

COMPANY: _____   PHONE: _____

TRUCK NO. _____   DRIVER: _____

TRAILER NO. _____   DRIVER: _____

DATE LEFT: _____   DATE RETURNED: _____

TRIP FROM: **TRIP TO:**

DATE	STATE	ODOMETER BEGINNING MILES	ODOMETER ENDING MILES	HIGHWAYS USED EACH STATE	STATE MILES

FUEL PURCHASE RECORD

DATE	NUMBER	GALLONS	COST	NAME	CITY	STATE	CASH/CHARGE CREDIT CARD

🚚 TRUCKER LOG

COMPANY: _____   PHONE: _____

TRUCK NO. _____   DRIVER: _____

TRAILER NO. _____   DRIVER: _____

DATE LEFT: _____   DATE RETURNED: _____

TRIP FROM: TRIP TO:

DATE	STATE	ODOMETER BEGINNING MILES	ODOMETER ENDING MILES	HIGHWAYS USED EACH STATE	STATE MILES

FUEL PURCHASE RECORD

DATE	NUMBER	GALLONS	COST	NAME	CITY	STATE	CASH/CHARGE CREDIT CARD

TRUCKER LOG

COMPANY: _____   PHONE: _____

TRUCK NO. _____   DRIVER: _____

TRAILER NO. _____   DRIVER: _____

DATE LEFT: _____   DATE RETURNED: _____

TRIP FROM: **TRIP TO:**

DATE	STATE	ODOMETER BEGINNING MILES	ODOMETER ENDING MILES	HIGHWAYS USED EACH STATE	STATE MILES

FUEL PURCHASE RECORD

DATE	NUMBER	GALLONS	COST	NAME	CITY	STATE	CASH/CHARGE CREDIT CARD

🚚 TRUCKER LOG

COMPANY: _____   PHONE: _____

TRUCK NO. _____   DRIVER: _____

TRAILER NO. _____   DRIVER: _____

DATE LEFT: _____   DATE RETURNED: _____

TRIP FROM: TRIP TO:

DATE	STATE	ODOMETER BEGINNING MILES	ODOMETER ENDING MILES	HIGHWAYS USED EACH STATE	STATE MILES

FUEL PURCHASE RECORD

DATE	NUMBER	GALLONS	COST	NAME	CITY	STATE	CASH/CHARGE CREDIT CARD

🚚 TRUCKER LOG

COMPANY: _____   PHONE: _____

TRUCK NO. _____   DRIVER: _____

TRAILER NO. _____   DRIVER: _____

DATE LEFT: _____   DATE RETURNED: _____

TRIP FROM: **TRIP TO:**

DATE	STATE	ODOMETER BEGINNING MILES	ODOMETER ENDING MILES	HIGHWAYS USED EACH STATE	STATE MILES

FUEL PURCHASE RECORD

DATE	NUMBER	GALLONS	COST	NAME	CITY	STATE	CASH/CHARGE CREDIT CARD

🚚 TRUCKER LOG

COMPANY: _____   PHONE: _____

TRUCK NO. _____   DRIVER: _____

TRAILER NO. _____   DRIVER: _____

DATE LEFT: _____   DATE RETURNED: _____

TRIP FROM: TRIP TO:

DATE	STATE	ODOMETER BEGINNING MILES	ODOMETER ENDING MILES	HIGHWAYS USED EACH STATE	STATE MILES

FUEL PURCHASE RECORD

DATE	NUMBER	GALLONS	COST	NAME	CITY	STATE	CASH/CHARGE CREDIT CARD

🚛 TRUCKER LOG

COMPANY: _____   PHONE: _____

TRUCK NO. _____   DRIVER: _____

TRAILER NO. _____   DRIVER: _____

DATE LEFT: _____   DATE RETURNED: _____

TRIP FROM: **TRIP TO:**

DATE	STATE	ODOMETER BEGINNING MILES	ODOMETER ENDING MILES	HIGHWAYS USED EACH STATE	STATE MILES

FUEL PURCHASE RECORD

DATE	NUMBER	GALLONS	COST	NAME	CITY	STATE	CASH/CHARGE CREDIT CARD

🚛 TRUCKER LOG

COMPANY: _____   PHONE: _____

TRUCK NO. _____   DRIVER: _____

TRAILER NO. _____   DRIVER: _____

DATE LEFT: _____   DATE RETURNED: _____

TRIP FROM: TRIP TO:

DATE	STATE	ODOMETER BEGINNING MILES	ODOMETER ENDING MILES	HIGHWAYS USED EACH STATE	STATE MILES

FUEL PURCHASE RECORD

DATE	NUMBER	GALLONS	COST	NAME	CITY	STATE	CASH/CHARGE CREDIT CARD

TRUCKER LOG

COMPANY: _____   PHONE: _____

TRUCK NO. _____   DRIVER: _____

TRAILER NO. _____   DRIVER: _____

DATE LEFT: _____   DATE RETURNED: _____

TRIP FROM: **TRIP TO:**

DATE	STATE	ODOMETER BEGINNING MILES	ODOMETER ENDING MILES	HIGHWAYS USED EACH STATE	STATE MILES

FUEL PURCHASE RECORD

DATE	NUMBER	GALLONS	COST	NAME	CITY	STATE	CASH/CHARGE CREDIT CARD

🚛 TRUCKER LOG

COMPANY: _____   PHONE: _____

TRUCK NO. _____   DRIVER: _____

TRAILER NO. _____   DRIVER: _____

DATE LEFT: _____   DATE RETURNED: _____

TRIP FROM: TRIP TO:

DATE	STATE	ODOMETER BEGINNING MILES	ODOMETER ENDING MILES	HIGHWAYS USED EACH STATE	STATE MILES

FUEL PURCHASE RECORD

DATE	NUMBER	GALLONS	COST	NAME	CITY	STATE	CASH/CHARGE CREDIT CARD

TRUCKER LOG

COMPANY: _____   PHONE: _____

TRUCK NO. _____   DRIVER: _____

TRAILER NO. _____   DRIVER: _____

DATE LEFT: _____   DATE RETURNED: _____

TRIP FROM: **TRIP TO:**

DATE	STATE	ODOMETER BEGINNING MILES	ODOMETER ENDING MILES	HIGHWAYS USED EACH STATE	STATE MILES

FUEL PURCHASE RECORD

DATE	NUMBER	GALLONS	COST	NAME	CITY	STATE	CASH/CHARGE CREDIT CARD

TRUCKER LOG

COMPANY: _____   PHONE: _____

TRUCK NO. _____   DRIVER: _____

TRAILER NO. _____   DRIVER: _____

DATE LEFT: _____   DATE RETURNED: _____

TRIP FROM: TRIP TO:

DATE	STATE	ODOMETER BEGINNING MILES	ODOMETER ENDING MILES	HIGHWAYS USED EACH STATE	STATE MILES

FUEL PURCHASE RECORD

DATE	NUMBER	GALLONS	COST	NAME	CITY	STATE	CASH/CHARGE CREDIT CARD

🚛 TRUCKER LOG

COMPANY: _____ PHONE: _____

TRUCK NO. _____ DRIVER: _____

TRAILER NO. _____ DRIVER: _____

DATE LEFT: _____ DATE RETURNED: _____

TRIP FROM: **TRIP TO:**

DATE	STATE	ODOMETER BEGINNING MILES	ODOMETER ENDING MILES	HIGHWAYS USED EACH STATE	STATE MILES

FUEL PURCHASE RECORD

DATE	NUMBER	GALLONS	COST	NAME	CITY	STATE	CASH/CHARGE CREDIT CARD

TRUCKER LOG

COMPANY: _____   PHONE: _____

TRUCK NO. _____   DRIVER: _____

TRAILER NO. _____   DRIVER: _____

DATE LEFT: _____   DATE RETURNED: _____

TRIP FROM: TRIP TO:

DATE	STATE	ODOMETER BEGINNING MILES	ODOMETER ENDING MILES	HIGHWAYS USED EACH STATE	STATE MILES

FUEL PURCHASE RECORD

DATE	NUMBER	GALLONS	COST	NAME	CITY	STATE	CASH/CHARGE CREDIT CARD

TRUCKER LOG

COMPANY: _____   PHONE: _____

TRUCK NO. _____   DRIVER: _____

TRAILER NO. _____   DRIVER: _____

DATE LEFT: _____   DATE RETURNED: _____

TRIP FROM: **TRIP TO:**

DATE	STATE	ODOMETER BEGINNING MILES	ODOMETER ENDING MILES	HIGHWAYS USED EACH STATE	STATE MILES

FUEL PURCHASE RECORD

DATE	NUMBER	GALLONS	COST	NAME	CITY	STATE	CASH/CHARGE CREDIT CARD

🚛 TRUCKER LOG

COMPANY: _____   PHONE: _____

TRUCK NO. _____   DRIVER: _____

TRAILER NO. _____   DRIVER: _____

DATE LEFT: _____   DATE RETURNED: _____

TRIP FROM: TRIP TO:

DATE	STATE	ODOMETER BEGINNING MILES	ODOMETER ENDING MILES	HIGHWAYS USED EACH STATE	STATE MILES

FUEL PURCHASE RECORD

DATE	NUMBER	GALLONS	COST	NAME	CITY	STATE	CASH/CHARGE CREDIT CARD

TRUCKER LOG

COMPANY: _____   PHONE: _____

TRUCK NO. _____   DRIVER: _____

TRAILER NO. _____   DRIVER: _____

DATE LEFT: _____   DATE RETURNED: _____

TRIP FROM: **TRIP TO:**

DATE	STATE	ODOMETER BEGINNING MILES	ODOMETER ENDING MILES	HIGHWAYS USED EACH STATE	STATE MILES

FUEL PURCHASE RECORD

DATE	NUMBER	GALLONS	COST	NAME	CITY	STATE	CASH/CHARGE CREDIT CARD

🚚 TRUCKER LOG

COMPANY: _____   PHONE: _____

TRUCK NO. _____   DRIVER: _____

TRAILER NO. _____   DRIVER: _____

DATE LEFT: _____   DATE RETURNED: _____

TRIP FROM: TRIP TO:

DATE	STATE	ODOMETER BEGINNING MILES	ODOMETER ENDING MILES	HIGHWAYS USED EACH STATE	STATE MILES

FUEL PURCHASE RECORD

DATE	NUMBER	GALLONS	COST	NAME	CITY	STATE	CASH/CHARGE CREDIT CARD

TRUCKER LOG

COMPANY: _____   PHONE: _____

TRUCK NO. _____   DRIVER: _____

TRAILER NO. _____   DRIVER: _____

DATE LEFT: _____   DATE RETURNED: _____

TRIP FROM: **TRIP TO:**

DATE	STATE	ODOMETER BEGINNING MILES	ODOMETER ENDING MILES	HIGHWAYS USED EACH STATE	STATE MILES

FUEL PURCHASE RECORD

DATE	NUMBER	GALLONS	COST	NAME	CITY	STATE	CASH/CHARGE CREDIT CARD

🚛 TRUCKER LOG

COMPANY: _____   PHONE: _____

TRUCK NO. _____   DRIVER: _____

TRAILER NO. _____   DRIVER: _____

DATE LEFT: _____   DATE RETURNED: _____

TRIP FROM: TRIP TO:

DATE	STATE	ODOMETER BEGINNING MILES	ODOMETER ENDING MILES	HIGHWAYS USED EACH STATE	STATE MILES

FUEL PURCHASE RECORD

DATE	NUMBER	GALLONS	COST	NAME	CITY	STATE	CASH/CHARGE CREDIT CARD

◢▬▬▬ TRUCKER LOG

COMPANY: _____   PHONE: _____

TRUCK NO. _____   DRIVER: _____

TRAILER NO. _____   DRIVER: _____

DATE LEFT: _____   DATE RETURNED: _____

TRIP FROM: **TRIP TO:**

DATE	STATE	ODOMETER BEGINNING MILES	ODOMETER ENDING MILES	HIGHWAYS USED EACH STATE	STATE MILES

FUEL PURCHASE RECORD

DATE	NUMBER	GALLONS	COST	NAME	CITY	STATE	CASH/CHARGE CREDIT CARD

TRUCKER LOG

COMPANY: _____ PHONE: _____

TRUCK NO. _____ DRIVER: _____

TRAILER NO. _____ DRIVER: _____

DATE LEFT: _____ DATE RETURNED: _____

TRIP FROM: **TRIP TO:**

DATE	STATE	ODOMETER BEGINNING MILES	ODOMETER ENDING MILES	HIGHWAYS USED EACH STATE	STATE MILES

FUEL PURCHASE RECORD

DATE	NUMBER	GALLONS	COST	NAME	CITY	STATE	CASH/CHARGE CREDIT CARD

TRUCKER LOG

COMPANY: _____   PHONE: _____

TRUCK NO. _____   DRIVER: _____

TRAILER NO. _____   DRIVER: _____

DATE LEFT: _____   DATE RETURNED: _____

TRIP FROM: **TRIP TO:**

DATE	STATE	ODOMETER BEGINNING MILES	ODOMETER ENDING MILES	HIGHWAYS USED EACH STATE	STATE MILES

FUEL PURCHASE RECORD

DATE	NUMBER	GALLONS	COST	NAME	CITY	STATE	CASH/CHARGE CREDIT CARD

🚚 TRUCKER LOG

COMPANY: _____   PHONE: _____

TRUCK NO. _____   DRIVER: _____

TRAILER NO. _____   DRIVER: _____

DATE LEFT: _____   DATE RETURNED: _____

TRIP FROM: TRIP TO:

DATE	STATE	ODOMETER BEGINNING MILES	ODOMETER ENDING MILES	HIGHWAYS USED EACH STATE	STATE MILES

FUEL PURCHASE RECORD

DATE	NUMBER	GALLONS	COST	NAME	CITY	STATE	CASH/CHARGE CREDIT CARD

TRUCKER LOG

COMPANY: _____   PHONE: _____

TRUCK NO. _____   DRIVER: _____

TRAILER NO. _____   DRIVER: _____

DATE LEFT: _____   DATE RETURNED: _____

TRIP FROM: **TRIP TO:**

DATE	STATE	ODOMETER BEGINNING MILES	ODOMETER ENDING MILES	HIGHWAYS USED EACH STATE	STATE MILES

FUEL PURCHASE RECORD

DATE	NUMBER	GALLONS	COST	NAME	CITY	STATE	CASH/CHARGE CREDIT CARD

🚛 TRUCKER LOG

COMPANY: _____   PHONE: _____

TRUCK NO. _____   DRIVER: _____

TRAILER NO. _____   DRIVER: _____

DATE LEFT: _____   DATE RETURNED: _____

TRIP FROM: TRIP TO:

DATE	STATE	ODOMETER BEGINNING MILES	ODOMETER ENDING MILES	HIGHWAYS USED EACH STATE	STATE MILES

FUEL PURCHASE RECORD

DATE	NUMBER	GALLONS	COST	NAME	CITY	STATE	CASH/CHARGE CREDIT CARD

TRUCKER LOG

COMPANY: _____ PHONE: _____

TRUCK NO. _____ DRIVER: _____

TRAILER NO. _____ DRIVER: _____

DATE LEFT: _____ DATE RETURNED: _____

TRIP FROM: **TRIP TO:**

DATE	STATE	ODOMETER BEGINNING MILES	ODOMETER ENDING MILES	HIGHWAYS USED EACH STATE	STATE MILES

FUEL PURCHASE RECORD

DATE	NUMBER	GALLONS	COST	NAME	CITY	STATE	CASH/CHARGE CREDIT CARD

TRUCKER LOG

COMPANY: _____   PHONE: _____

TRUCK NO. _____   DRIVER: _____

TRAILER NO. _____   DRIVER: _____

DATE LEFT: _____   DATE RETURNED: _____

TRIP FROM: TRIP TO:

DATE	STATE	ODOMETER BEGINNING MILES	ODOMETER ENDING MILES	HIGHWAYS USED EACH STATE	STATE MILES

FUEL PURCHASE RECORD

DATE	NUMBER	GALLONS	COST	NAME	CITY	STATE	CASH/CHARGE CREDIT CARD

TRUCKER LOG

COMPANY: _____   PHONE: _____

TRUCK NO. _____   DRIVER: _____

TRAILER NO. _____   DRIVER: _____

DATE LEFT: _____   DATE RETURNED: _____

TRIP FROM: **TRIP TO:**

DATE	STATE	ODOMETER BEGINNING MILES	ODOMETER ENDING MILES	HIGHWAYS USED EACH STATE	STATE MILES

FUEL PURCHASE RECORD

DATE	NUMBER	GALLONS	COST	NAME	CITY	STATE	CASH/CHARGE CREDIT CARD

🚛 TRUCKER LOG

COMPANY: _____   PHONE: _____

TRUCK NO. _____   DRIVER: _____

TRAILER NO. _____   DRIVER: _____

DATE LEFT: _____   DATE RETURNED: _____

TRIP FROM: TRIP TO:

DATE	STATE	ODOMETER BEGINNING MILES	ODOMETER ENDING MILES	HIGHWAYS USED EACH STATE	STATE MILES

FUEL PURCHASE RECORD

DATE	NUMBER	GALLONS	COST	NAME	CITY	STATE	CASH/CHARGE CREDIT CARD

TRUCKER LOG

COMPANY: _____   PHONE: _____

TRUCK NO. _____   DRIVER: _____

TRAILER NO. _____   DRIVER: _____

DATE LEFT: _____   DATE RETURNED: _____

TRIP FROM: TRIP TO:

DATE	STATE	ODOMETER BEGINNING MILES	ODOMETER ENDING MILES	HIGHWAYS USED EACH STATE	STATE MILES

FUEL PURCHASE RECORD

DATE	NUMBER	GALLONS	COST	NAME	CITY	STATE	CASH/CHARGE CREDIT CARD

🚚 TRUCKER LOG

COMPANY: _____   PHONE: _____

TRUCK NO. _____   DRIVER: _____

TRAILER NO. _____   DRIVER: _____

DATE LEFT: _____   DATE RETURNED: _____

TRIP FROM: TRIP TO:

DATE	STATE	ODOMETER BEGINNING MILES	ODOMETER ENDING MILES	HIGHWAYS USED EACH STATE	STATE MILES

FUEL PURCHASE RECORD

DATE	NUMBER	GALLONS	COST	NAME	CITY	STATE	CASH/CHARGE CREDIT CARD

🚛 TRUCKER LOG

COMPANY: _____   PHONE: _____

TRUCK NO. _____   DRIVER: _____

TRAILER NO. _____   DRIVER: _____

DATE LEFT: _____   DATE RETURNED: _____

TRIP FROM: **TRIP TO:**

DATE	STATE	ODOMETER BEGINNING MILES	ODOMETER ENDING MILES	HIGHWAYS USED EACH STATE	STATE MILES

FUEL PURCHASE RECORD

DATE	NUMBER	GALLONS	COST	NAME	CITY	STATE	CASH/CHARGE CREDIT CARD

🚚 TRUCKER LOG

COMPANY: _____   PHONE: _____

TRUCK NO. _____   DRIVER: _____

TRAILER NO. _____   DRIVER: _____

DATE LEFT: _____   DATE RETURNED: _____

TRIP FROM: TRIP TO:

DATE	STATE	ODOMETER BEGINNING MILES	ODOMETER ENDING MILES	HIGHWAYS USED EACH STATE	STATE MILES

FUEL PURCHASE RECORD

DATE	NUMBER	GALLONS	COST	NAME	CITY	STATE	CASH/CHARGE CREDIT CARD

TRUCKER LOG

COMPANY: _____   PHONE: _____

TRUCK NO. _____   DRIVER: _____

TRAILER NO. _____   DRIVER: _____

DATE LEFT: _____   DATE RETURNED: _____

TRIP FROM: TRIP TO:

DATE	STATE	ODOMETER BEGINNING MILES	ODOMETER ENDING MILES	HIGHWAYS USED EACH STATE	STATE MILES

FUEL PURCHASE RECORD

DATE	NUMBER	GALLONS	COST	NAME	CITY	STATE	CASH/CHARGE CREDIT CARD

🚛 TRUCKER LOG

COMPANY: _____   PHONE: _____

TRUCK NO. _____   DRIVER: _____

TRAILER NO. _____   DRIVER: _____

DATE LEFT: _____   DATE RETURNED: _____

TRIP FROM: TRIP TO:

DATE	STATE	ODOMETER BEGINNING MILES	ODOMETER ENDING MILES	HIGHWAYS USED EACH STATE	STATE MILES

FUEL PURCHASE RECORD

DATE	NUMBER	GALLONS	COST	NAME	CITY	STATE	CASH/CHARGE CREDIT CARD

🚛 TRUCKER LOG

COMPANY: _____   PHONE: _____

TRUCK NO. _____   DRIVER: _____

TRAILER NO. _____   DRIVER: _____

DATE LEFT: _____   DATE RETURNED: _____

TRIP FROM: **TRIP TO:**

DATE	STATE	ODOMETER BEGINNING MILES	ODOMETER ENDING MILES	HIGHWAYS USED EACH STATE	STATE MILES

FUEL PURCHASE RECORD

DATE	NUMBER	GALLONS	COST	NAME	CITY	STATE	CASH/CHARGE CREDIT CARD

TRUCKER LOG

COMPANY: _____   PHONE: _____

TRUCK NO. _____   DRIVER: _____

TRAILER NO. _____   DRIVER: _____

DATE LEFT: _____   DATE RETURNED: _____

TRIP FROM: **TRIP TO:**

DATE	STATE	ODOMETER BEGINNING MILES	ODOMETER ENDING MILES	HIGHWAYS USED EACH STATE	STATE MILES

FUEL PURCHASE RECORD

DATE	NUMBER	GALLONS	COST	NAME	CITY	STATE	CASH/CHARGE CREDIT CARD

TRUCKER LOG

COMPANY: _____

TRUCK NO. _____

TRAILER NO. _____

DATE LEFT: _____

PHONE: _____

DRIVER: _____

DRIVER: _____

DATE RETURNED: _____

TRIP FROM:

TRIP TO:

DATE	STATE	ODOMETER BEGINNING MILES	ODOMETER ENDING MILES	HIGHWAYS USED EACH STATE	STATE MILES

FUEL PURCHASE RECORD

DATE	NUMBER	GALLONS	COST	NAME	CITY	STATE	CASH/CHARGE CREDIT CARD

🚚 TRUCKER LOG

COMPANY: _____ PHONE: _____

TRUCK NO. _____ DRIVER: _____

TRAILER NO. _____ DRIVER: _____

DATE LEFT: _____ DATE RETURNED: _____

TRIP FROM: | **TRIP TO:**

DATE	STATE	ODOMETER BEGINNING MILES	ODOMETER ENDING MILES	HIGHWAYS USED EACH STATE	STATE MILES

FUEL PURCHASE RECORD

DATE	NUMBER	GALLONS	COST	NAME	CITY	STATE	CASH/CHARGE CREDIT CARD

TRUCKER LOG

COMPANY: _____   PHONE: _____

TRUCK NO. _____   DRIVER: _____

TRAILER NO. _____   DRIVER: _____

DATE LEFT: _____   DATE RETURNED: _____

TRIP FROM: **TRIP TO:**

DATE	STATE	ODOMETER BEGINNING MILES	ODOMETER ENDING MILES	HIGHWAYS USED EACH STATE	STATE MILES

FUEL PURCHASE RECORD

DATE	NUMBER	GALLONS	COST	NAME	CITY	STATE	CASH/CHARGE CREDIT CARD

🚚 TRUCKER LOG

COMPANY: _____   PHONE: _____

TRUCK NO. _____   DRIVER: _____

TRAILER NO. _____   DRIVER: _____

DATE LEFT: _____   DATE RETURNED: _____

TRIP FROM: TRIP TO:

DATE	STATE	ODOMETER BEGINNING MILES	ODOMETER ENDING MILES	HIGHWAYS USED EACH STATE	STATE MILES

FUEL PURCHASE RECORD

DATE	NUMBER	GALLONS	COST	NAME	CITY	STATE	CASH/CHARGE CREDIT CARD

◢▰▰▰▰ TRUCKER LOG

COMPANY: _____ PHONE: _____

TRUCK NO. _____ DRIVER: _____

TRAILER NO. _____ DRIVER: _____

DATE LEFT: _____ DATE RETURNED: _____

TRIP FROM: **TRIP TO:**

DATE	STATE	ODOMETER BEGINNING MILES	ODOMETER ENDING MILES	HIGHWAYS USED EACH STATE	STATE MILES

FUEL PURCHASE RECORD

DATE	NUMBER	GALLONS	COST	NAME	CITY	STATE	CASH/CHARGE CREDIT CARD

◀▬▬▬ TRUCKER LOG

COMPANY: _____   PHONE: _____

TRUCK NO. _____   DRIVER: _____

TRAILER NO. _____   DRIVER: _____

DATE LEFT: _____   DATE RETURNED: _____

TRIP FROM: TRIP TO:

DATE	STATE	ODOMETER BEGINNING MILES	ODOMETER ENDING MILES	HIGHWAYS USED EACH STATE	STATE MILES

FUEL PURCHASE RECORD

DATE	NUMBER	GALLONS	COST	NAME	CITY	STATE	CASH/CHARGE CREDIT CARD

TRUCKER LOG

COMPANY: _____ PHONE: _____

TRUCK NO. _____ DRIVER: _____

TRAILER NO. _____ DRIVER: _____

DATE LEFT: _____ DATE RETURNED: _____

TRIP FROM: **TRIP TO:**

DATE	STATE	ODOMETER BEGINNING MILES	ODOMETER ENDING MILES	HIGHWAYS USED EACH STATE	STATE MILES

FUEL PURCHASE RECORD

DATE	NUMBER	GALLONS	COST	NAME	CITY	STATE	CASH/CHARGE CREDIT CARD

🚛 TRUCKER LOG

COMPANY: _____   PHONE: _____

TRUCK NO. _____   DRIVER: _____

TRAILER NO. _____   DRIVER: _____

DATE LEFT: _____   DATE RETURNED: _____

TRIP FROM: TRIP TO:

DATE	STATE	ODOMETER BEGINNING MILES	ODOMETER ENDING MILES	HIGHWAYS USED EACH STATE	STATE MILES

FUEL PURCHASE RECORD

DATE	NUMBER	GALLONS	COST	NAME	CITY	STATE	CASH/CHARGE CREDIT CARD

TRUCKER LOG

COMPANY: _____   PHONE: _____

TRUCK NO. _____   DRIVER: _____

TRAILER NO. _____   DRIVER: _____

DATE LEFT: _____   DATE RETURNED: _____

TRIP FROM: TRIP TO:

DATE	STATE	ODOMETER BEGINNING MILES	ODOMETER ENDING MILES	HIGHWAYS USED EACH STATE	STATE MILES

FUEL PURCHASE RECORD

DATE	NUMBER	GALLONS	COST	NAME	CITY	STATE	CASH/CHARGE CREDIT CARD

TRUCKER LOG

COMPANY: _____   PHONE: _____

TRUCK NO. _____   DRIVER: _____

TRAILER NO. _____   DRIVER: _____

DATE LEFT: _____   DATE RETURNED: _____

TRIP FROM: TRIP TO:

DATE	STATE	ODOMETER BEGINNING MILES	ODOMETER ENDING MILES	HIGHWAYS USED EACH STATE	STATE MILES

FUEL PURCHASE RECORD

DATE	NUMBER	GALLONS	COST	NAME	CITY	STATE	CASH/CHARGE CREDIT CARD

TRUCKER LOG

COMPANY: _____ PHONE: _____

TRUCK NO. _____ DRIVER: _____

TRAILER NO. _____ DRIVER: _____

DATE LEFT: _____ DATE RETURNED: _____

TRIP FROM: **TRIP TO:**

DATE	STATE	ODOMETER BEGINNING MILES	ODOMETER ENDING MILES	HIGHWAYS USED EACH STATE	STATE MILES

FUEL PURCHASE RECORD

DATE	NUMBER	GALLONS	COST	NAME	CITY	STATE	CASH/CHARGE CREDIT CARD

TRUCKER LOG

COMPANY: _____   PHONE: _____

TRUCK NO. _____   DRIVER: _____

TRAILER NO. _____   DRIVER: _____

DATE LEFT: _____   DATE RETURNED: _____

TRIP FROM: TRIP TO:

DATE	STATE	ODOMETER BEGINNING MILES	ODOMETER ENDING MILES	HIGHWAYS USED EACH STATE	STATE MILES

FUEL PURCHASE RECORD

DATE	NUMBER	GALLONS	COST	NAME	CITY	STATE	CASH/CHARGE CREDIT CARD

TRUCKER LOG

COMPANY: _____   PHONE: _____

TRUCK NO. _____   DRIVER: _____

TRAILER NO. _____   DRIVER: _____

DATE LEFT: _____   DATE RETURNED: _____

TRIP FROM: **TRIP TO:**

DATE	STATE	ODOMETER BEGINNING MILES	ODOMETER ENDING MILES	HIGHWAYS USED EACH STATE	STATE MILES

FUEL PURCHASE RECORD

DATE	NUMBER	GALLONS	COST	NAME	CITY	STATE	CASH/CHARGE CREDIT CARD

🚚 TRUCKER LOG

COMPANY: _____   PHONE: _____

TRUCK NO. _____   DRIVER: _____

TRAILER NO. _____   DRIVER: _____

DATE LEFT: _____   DATE RETURNED: _____

TRIP FROM: TRIP TO:

DATE	STATE	ODOMETER BEGINNING MILES	ODOMETER ENDING MILES	HIGHWAYS USED EACH STATE	STATE MILES

FUEL PURCHASE RECORD

DATE	NUMBER	GALLONS	COST	NAME	CITY	STATE	CASH/CHARGE CREDIT CARD

TRUCKER LOG

COMPANY: _____   PHONE: _____

TRUCK NO. _____   DRIVER: _____

TRAILER NO. _____   DRIVER: _____

DATE LEFT: _____   DATE RETURNED: _____

TRIP FROM: **TRIP TO:**

DATE	STATE	ODOMETER BEGINNING MILES	ODOMETER ENDING MILES	HIGHWAYS USED EACH STATE	STATE MILES

FUEL PURCHASE RECORD

DATE	NUMBER	GALLONS	COST	NAME	CITY	STATE	CASH/CHARGE CREDIT CARD

◢▬▬▬ TRUCKER LOG

COMPANY: _____   PHONE: _____

TRUCK NO. _____   DRIVER: _____

TRAILER NO. _____   DRIVER: _____

DATE LEFT: _____   DATE RETURNED: _____

TRIP FROM: TRIP TO:

DATE	STATE	ODOMETER BEGINNING MILES	ODOMETER ENDING MILES	HIGHWAYS USED EACH STATE	STATE MILES

FUEL PURCHASE RECORD

DATE	NUMBER	GALLONS	COST	NAME	CITY	STATE	CASH/CHARGE CREDIT CARD

TRUCKER LOG

COMPANY: _____   PHONE: _____

TRUCK NO. _____   DRIVER: _____

TRAILER NO. _____   DRIVER: _____

DATE LEFT: _____   DATE RETURNED: _____

TRIP FROM: **TRIP TO:**

DATE	STATE	ODOMETER BEGINNING MILES	ODOMETER ENDING MILES	HIGHWAYS USED EACH STATE	STATE MILES

FUEL PURCHASE RECORD

DATE	NUMBER	GALLONS	COST	NAME	CITY	STATE	CASH/CHARGE CREDIT CARD

TRUCKER LOG

COMPANY: _____   PHONE: _____

TRUCK NO. _____   DRIVER: _____

TRAILER NO. _____   DRIVER: _____

DATE LEFT: _____   DATE RETURNED: _____

TRIP FROM: TRIP TO:

DATE	STATE	ODOMETER BEGINNING MILES	ODOMETER ENDING MILES	HIGHWAYS USED EACH STATE	STATE MILES

FUEL PURCHASE RECORD

DATE	NUMBER	GALLONS	COST	NAME	CITY	STATE	CASH/CHARGE CREDIT CARD

🚛 TRUCKER LOG

COMPANY: _____   PHONE: _____

TRUCK NO. _____   DRIVER: _____

TRAILER NO. _____   DRIVER: _____

DATE LEFT: _____   DATE RETURNED: _____

TRIP FROM: **TRIP TO:**

DATE	STATE	ODOMETER BEGINNING MILES	ODOMETER ENDING MILES	HIGHWAYS USED EACH STATE	STATE MILES

FUEL PURCHASE RECORD

DATE	NUMBER	GALLONS	COST	NAME	CITY	STATE	CASH/CHARGE CREDIT CARD

TRUCKER LOG

COMPANY: _____ PHONE: _____

TRUCK NO. _____ DRIVER: _____

TRAILER NO. _____ DRIVER: _____

DATE LEFT: _____ DATE RETURNED: _____

TRIP FROM: TRIP TO:

DATE	STATE	ODOMETER BEGINNING MILES	ODOMETER ENDING MILES	HIGHWAYS USED EACH STATE	STATE MILES

FUEL PURCHASE RECORD

DATE	NUMBER	GALLONS	COST	NAME	CITY	STATE	CASH/CHARGE CREDIT CARD

◢▬▬▬ TRUCKER LOG

COMPANY: _____   PHONE: _____

TRUCK NO. _____   DRIVER: _____

TRAILER NO. _____   DRIVER: _____

DATE LEFT: _____   DATE RETURNED: _____

TRIP FROM: **TRIP TO:**

DATE	STATE	ODOMETER BEGINNING MILES	ODOMETER ENDING MILES	HIGHWAYS USED EACH STATE	STATE MILES

FUEL PURCHASE RECORD

DATE	NUMBER	GALLONS	COST	NAME	CITY	STATE	CASH/CHARGE CREDIT CARD

🚚 TRUCKER LOG

COMPANY: _____ PHONE: _____

TRUCK NO. _____ DRIVER: _____

TRAILER NO. _____ DRIVER: _____

DATE LEFT: _____ DATE RETURNED: _____

TRIP FROM: TRIP TO:

DATE	STATE	ODOMETER BEGINNING MILES	ODOMETER ENDING MILES	HIGHWAYS USED EACH STATE	STATE MILES

FUEL PURCHASE RECORD

DATE	NUMBER	GALLONS	COST	NAME	CITY	STATE	CASH/CHARGE CREDIT CARD

TRUCKER LOG

COMPANY: _____   PHONE: _____

TRUCK NO. _____   DRIVER: _____

TRAILER NO. _____   DRIVER: _____

DATE LEFT: _____   DATE RETURNED: _____

TRIP FROM: **TRIP TO:**

DATE	STATE	ODOMETER BEGINNING MILES	ODOMETER ENDING MILES	HIGHWAYS USED EACH STATE	STATE MILES

FUEL PURCHASE RECORD

DATE	NUMBER	GALLONS	COST	NAME	CITY	STATE	CASH/CHARGE CREDIT CARD

TRUCKER LOG

COMPANY: _____   PHONE: _____

TRUCK NO. _____   DRIVER: _____

TRAILER NO. _____   DRIVER: _____

DATE LEFT: _____   DATE RETURNED: _____

TRIP FROM: TRIP TO:

DATE	STATE	ODOMETER BEGINNING MILES	ODOMETER ENDING MILES	HIGHWAYS USED EACH STATE	STATE MILES

FUEL PURCHASE RECORD

DATE	NUMBER	GALLONS	COST	NAME	CITY	STATE	CASH/CHARGE CREDIT CARD

TRUCKER LOG

COMPANY: _____   PHONE: _____

TRUCK NO. _____   DRIVER: _____

TRAILER NO. _____   DRIVER: _____

DATE LEFT: _____   DATE RETURNED: _____

TRIP FROM: **TRIP TO:**

DATE	STATE	ODOMETER BEGINNING MILES	ODOMETER ENDING MILES	HIGHWAYS USED EACH STATE	STATE MILES

FUEL PURCHASE RECORD

DATE	NUMBER	GALLONS	COST	NAME	CITY	STATE	CASH/CHARGE CREDIT CARD

TRUCKER LOG

COMPANY: _____   PHONE: _____

TRUCK NO. _____   DRIVER: _____

TRAILER NO. _____   DRIVER: _____

DATE LEFT: _____   DATE RETURNED: _____

TRIP FROM: TRIP TO:

DATE	STATE	ODOMETER BEGINNING MILES	ODOMETER ENDING MILES	HIGHWAYS USED EACH STATE	STATE MILES

FUEL PURCHASE RECORD

DATE	NUMBER	GALLONS	COST	NAME	CITY	STATE	CASH/CHARGE CREDIT CARD

🚚 TRUCKER LOG

COMPANY: _____   PHONE: _____

TRUCK NO. _____   DRIVER: _____

TRAILER NO. _____   DRIVER: _____

DATE LEFT: _____   DATE RETURNED: _____

TRIP FROM: **TRIP TO:**

DATE	STATE	ODOMETER BEGINNING MILES	ODOMETER ENDING MILES	HIGHWAYS USED EACH STATE	STATE MILES

FUEL PURCHASE RECORD

DATE	NUMBER	GALLONS	COST	NAME	CITY	STATE	CASH/CHARGE CREDIT CARD

▬▬ TRUCKER LOG

COMPANY: _____   PHONE: _____

TRUCK NO. _____   DRIVER: _____

TRAILER NO. _____   DRIVER: _____

DATE LEFT: _____   DATE RETURNED: _____

TRIP FROM: TRIP TO:

DATE	STATE	ODOMETER BEGINNING MILES	ODOMETER ENDING MILES	HIGHWAYS USED EACH STATE	STATE MILES

FUEL PURCHASE RECORD

DATE	NUMBER	GALLONS	COST	NAME	CITY	STATE	CASH/CHARGE CREDIT CARD

TRUCKER LOG

COMPANY: _____ PHONE: _____

TRUCK NO. _____ DRIVER: _____

TRAILER NO. _____ DRIVER: _____

DATE LEFT: _____ DATE RETURNED: _____

TRIP FROM: **TRIP TO:**

DATE	STATE	ODOMETER BEGINNING MILES	ODOMETER ENDING MILES	HIGHWAYS USED EACH STATE	STATE MILES

FUEL PURCHASE RECORD

DATE	NUMBER	GALLONS	COST	NAME	CITY	STATE	CASH/CHARGE CREDIT CARD

TRUCKER LOG

COMPANY: _____   PHONE: _____

TRUCK NO. _____   DRIVER: _____

TRAILER NO. _____   DRIVER: _____

DATE LEFT: _____   DATE RETURNED: _____

TRIP FROM: TRIP TO:

DATE	STATE	ODOMETER BEGINNING MILES	ODOMETER ENDING MILES	HIGHWAYS USED EACH STATE	STATE MILES

FUEL PURCHASE RECORD

DATE	NUMBER	GALLONS	COST	NAME	CITY	STATE	CASH/CHARGE CREDIT CARD

🚚 TRUCKER LOG

COMPANY: _____   PHONE: _____

TRUCK NO. _____   DRIVER: _____

TRAILER NO. _____   DRIVER: _____

DATE LEFT: _____   DATE RETURNED: _____

TRIP FROM: **TRIP TO:**

DATE	STATE	ODOMETER BEGINNING MILES	ODOMETER ENDING MILES	HIGHWAYS USED EACH STATE	STATE MILES

FUEL PURCHASE RECORD

DATE	NUMBER	GALLONS	COST	NAME	CITY	STATE	CASH/CHARGE CREDIT CARD

TRUCKER LOG

COMPANY: _____   PHONE: _____

TRUCK NO. _____   DRIVER: _____

TRAILER NO. _____   DRIVER: _____

DATE LEFT: _____   DATE RETURNED: _____

TRIP FROM: **TRIP TO:**

DATE	STATE	ODOMETER BEGINNING MILES	ODOMETER ENDING MILES	HIGHWAYS USED EACH STATE	STATE MILES

FUEL PURCHASE RECORD

DATE	NUMBER	GALLONS	COST	NAME	CITY	STATE	CASH/CHARGE CREDIT CARD

TRUCKER LOG

COMPANY: _____   PHONE: _____

TRUCK NO. _____   DRIVER: _____

TRAILER NO. _____   DRIVER: _____

DATE LEFT: _____   DATE RETURNED: _____

TRIP FROM: **TRIP TO:**

DATE	STATE	ODOMETER BEGINNING MILES	ODOMETER ENDING MILES	HIGHWAYS USED EACH STATE	STATE MILES

FUEL PURCHASE RECORD

DATE	NUMBER	GALLONS	COST	NAME	CITY	STATE	CASH/CHARGE CREDIT CARD

TRUCKER LOG

COMPANY: _____   PHONE: _____

TRUCK NO. _____   DRIVER: _____

TRAILER NO. _____   DRIVER: _____

DATE LEFT: _____   DATE RETURNED: _____

TRIP FROM: TRIP TO:

DATE	STATE	ODOMETER BEGINNING MILES	ODOMETER ENDING MILES	HIGHWAYS USED EACH STATE	STATE MILES

FUEL PURCHASE RECORD

DATE	NUMBER	GALLONS	COST	NAME	CITY	STATE	CASH/CHARGE CREDIT CARD

TRUCKER LOG

COMPANY: _____   PHONE: _____

TRUCK NO. _____   DRIVER: _____

TRAILER NO. _____   DRIVER: _____

DATE LEFT: _____   DATE RETURNED: _____

TRIP FROM: **TRIP TO:**

DATE	STATE	ODOMETER BEGINNING MILES	ODOMETER ENDING MILES	HIGHWAYS USED EACH STATE	STATE MILES

FUEL PURCHASE RECORD

DATE	NUMBER	GALLONS	COST	NAME	CITY	STATE	CASH/CHARGE CREDIT CARD

🚚 TRUCKER LOG

COMPANY: _____   PHONE: _____

TRUCK NO. _____   DRIVER: _____

TRAILER NO. _____   DRIVER: _____

DATE LEFT: _____   DATE RETURNED: _____

TRIP FROM: TRIP TO:

DATE	STATE	ODOMETER BEGINNING MILES	ODOMETER ENDING MILES	HIGHWAYS USED EACH STATE	STATE MILES

FUEL PURCHASE RECORD

DATE	NUMBER	GALLONS	COST	NAME	CITY	STATE	CASH/CHARGE CREDIT CARD

TRUCKER LOG

COMPANY: _____ PHONE: _____

TRUCK NO. _____ DRIVER: _____

TRAILER NO. _____ DRIVER: _____

DATE LEFT: _____ DATE RETURNED: _____

TRIP FROM: **TRIP TO:**

DATE	STATE	ODOMETER BEGINNING MILES	ODOMETER ENDING MILES	HIGHWAYS USED EACH STATE	STATE MILES

FUEL PURCHASE RECORD

DATE	NUMBER	GALLONS	COST	NAME	CITY	STATE	CASH/CHARGE CREDIT CARD

TRUCKER LOG

COMPANY: _____   PHONE: _____

TRUCK NO. _____   DRIVER: _____

TRAILER NO. _____   DRIVER: _____

DATE LEFT: _____   DATE RETURNED: _____

TRIP FROM: TRIP TO:

DATE	STATE	ODOMETER BEGINNING MILES	ODOMETER ENDING MILES	HIGHWAYS USED EACH STATE	STATE MILES

FUEL PURCHASE RECORD

DATE	NUMBER	GALLONS	COST	NAME	CITY	STATE	CASH/CHARGE CREDIT CARD

TRUCKER LOG

COMPANY: _____ PHONE: _____

TRUCK NO. _____ DRIVER: _____

TRAILER NO. _____ DRIVER: _____

DATE LEFT: _____ DATE RETURNED: _____

TRIP FROM: TRIP TO:

DATE	STATE	ODOMETER BEGINNING MILES	ODOMETER ENDING MILES	HIGHWAYS USED EACH STATE	STATE MILES

FUEL PURCHASE RECORD

DATE	NUMBER	GALLONS	COST	NAME	CITY	STATE	CASH/CHARGE CREDIT CARD

🚚 TRUCKER LOG

COMPANY: _____   PHONE: _____

TRUCK NO. _____   DRIVER: _____

TRAILER NO. _____   DRIVER: _____

DATE LEFT: _____   DATE RETURNED: _____

TRIP FROM: TRIP TO:

DATE	STATE	ODOMETER BEGINNING MILES	ODOMETER ENDING MILES	HIGHWAYS USED EACH STATE	STATE MILES

FUEL PURCHASE RECORD

DATE	NUMBER	GALLONS	COST	NAME	CITY	STATE	CASH/CHARGE CREDIT CARD

◢◼◼◼ TRUCKER LOG

COMPANY: _____ PHONE: _____

TRUCK NO. _____ DRIVER: _____

TRAILER NO. _____ DRIVER: _____

DATE LEFT: _____ DATE RETURNED: _____

TRIP FROM: **TRIP TO:**

DATE	STATE	ODOMETER BEGINNING MILES	ODOMETER ENDING MILES	HIGHWAYS USED EACH STATE	STATE MILES

FUEL PURCHASE RECORD

DATE	NUMBER	GALLONS	COST	NAME	CITY	STATE	CASH/CHARGE CREDIT CARD

TRUCKER LOG

COMPANY: _____   PHONE: _____

TRUCK NO. _____   DRIVER: _____

TRAILER NO. _____   DRIVER: _____

DATE LEFT: _____   DATE RETURNED: _____

TRIP FROM: TRIP TO:

DATE	STATE	ODOMETER BEGINNING MILES	ODOMETER ENDING MILES	HIGHWAYS USED EACH STATE	STATE MILES

FUEL PURCHASE RECORD

DATE	NUMBER	GALLONS	COST	NAME	CITY	STATE	CASH/CHARGE CREDIT CARD

TRUCKER LOG

COMPANY: _____ PHONE: _____

TRUCK NO. _____ DRIVER: _____

TRAILER NO. _____ DRIVER: _____

DATE LEFT: _____ DATE RETURNED: _____

TRIP FROM: **TRIP TO:**

DATE	STATE	ODOMETER BEGINNING MILES	ODOMETER ENDING MILES	HIGHWAYS USED EACH STATE	STATE MILES

FUEL PURCHASE RECORD

DATE	NUMBER	GALLONS	COST	NAME	CITY	STATE	CASH/CHARGE CREDIT CARD

🚚 TRUCKER LOG

COMPANY: _____ PHONE: _____

TRUCK NO. _____ DRIVER: _____

TRAILER NO. _____ DRIVER: _____

DATE LEFT: _____ DATE RETURNED: _____

TRIP FROM: TRIP TO:

DATE	STATE	ODOMETER BEGINNING MILES	ODOMETER ENDING MILES	HIGHWAYS USED EACH STATE	STATE MILES

FUEL PURCHASE RECORD

DATE	NUMBER	GALLONS	COST	NAME	CITY	STATE	CASH/CHARGE CREDIT CARD

🚚 TRUCKER LOG

COMPANY: _____   PHONE: _____

TRUCK NO. _____   DRIVER: _____

TRAILER NO. _____   DRIVER: _____

DATE LEFT: _____   DATE RETURNED: _____

TRIP FROM: **TRIP TO:**

DATE	STATE	ODOMETER BEGINNING MILES	ODOMETER ENDING MILES	HIGHWAYS USED EACH STATE	STATE MILES

FUEL PURCHASE RECORD

DATE	NUMBER	GALLONS	COST	NAME	CITY	STATE	CASH/CHARGE CREDIT CARD

🚛 TRUCKER LOG

COMPANY: _____   PHONE: _____

TRUCK NO. _____   DRIVER: _____

TRAILER NO. _____   DRIVER: _____

DATE LEFT: _____   DATE RETURNED: _____

TRIP FROM: TRIP TO:

DATE	STATE	ODOMETER BEGINNING MILES	ODOMETER ENDING MILES	HIGHWAYS USED EACH STATE	STATE MILES

FUEL PURCHASE RECORD

DATE	NUMBER	GALLONS	COST	NAME	CITY	STATE	CASH/CHARGE CREDIT CARD

TRUCKER LOG

COMPANY: _____   PHONE: _____

TRUCK NO. _____   DRIVER: _____

TRAILER NO. _____   DRIVER: _____

DATE LEFT: _____   DATE RETURNED: _____

TRIP FROM: **TRIP TO:**

DATE	STATE	ODOMETER BEGINNING MILES	ODOMETER ENDING MILES	HIGHWAYS USED EACH STATE	STATE MILES

FUEL PURCHASE RECORD

DATE	NUMBER	GALLONS	COST	NAME	CITY	STATE	CASH/CHARGE CREDIT CARD

TRUCKER LOG

COMPANY: _____   PHONE: _____

TRUCK NO. _____   DRIVER: _____

TRAILER NO. _____   DRIVER: _____

DATE LEFT: _____   DATE RETURNED: _____

TRIP FROM: TRIP TO:

DATE	STATE	ODOMETER BEGINNING MILES	ODOMETER ENDING MILES	HIGHWAYS USED EACH STATE	STATE MILES

FUEL PURCHASE RECORD

DATE	NUMBER	GALLONS	COST	NAME	CITY	STATE	CASH/CHARGE CREDIT CARD

🚚 TRUCKER LOG

COMPANY: _____   PHONE: _____

TRUCK NO. _____   DRIVER: _____

TRAILER NO. _____   DRIVER: _____

DATE LEFT: _____   DATE RETURNED: _____

TRIP FROM: **TRIP TO:**

DATE	STATE	ODOMETER BEGINNING MILES	ODOMETER ENDING MILES	HIGHWAYS USED EACH STATE	STATE MILES

FUEL PURCHASE RECORD

DATE	NUMBER	GALLONS	COST	NAME	CITY	STATE	CASH/CHARGE CREDIT CARD

TRUCKER LOG

COMPANY: _____   PHONE: _____

TRUCK NO. _____   DRIVER: _____

TRAILER NO. _____   DRIVER: _____

DATE LEFT: _____   DATE RETURNED: _____

TRIP FROM: **TRIP TO:**

DATE	STATE	ODOMETER BEGINNING MILES	ODOMETER ENDING MILES	HIGHWAYS USED EACH STATE	STATE MILES

FUEL PURCHASE RECORD

DATE	NUMBER	GALLONS	COST	NAME	CITY	STATE	CASH/CHARGE CREDIT CARD

TRUCKER LOG

COMPANY: _____ PHONE: _____

TRUCK NO. _____ DRIVER: _____

TRAILER NO. _____ DRIVER: _____

DATE LEFT: _____ DATE RETURNED: _____

TRIP FROM: TRIP TO:

DATE	STATE	ODOMETER BEGINNING MILES	ODOMETER ENDING MILES	HIGHWAYS USED EACH STATE	STATE MILES

FUEL PURCHASE RECORD

DATE	NUMBER	GALLONS	COST	NAME	CITY	STATE	CASH/CHARGE CREDIT CARD

TRUCKER LOG

COMPANY: _____   PHONE: _____

TRUCK NO. _____   DRIVER: _____

TRAILER NO. _____   DRIVER: _____

DATE LEFT: _____   DATE RETURNED: _____

TRIP FROM: TRIP TO:

DATE	STATE	ODOMETER BEGINNING MILES	ODOMETER ENDING MILES	HIGHWAYS USED EACH STATE	STATE MILES

FUEL PURCHASE RECORD

DATE	NUMBER	GALLONS	COST	NAME	CITY	STATE	CASH/CHARGE CREDIT CARD

TRUCKER LOG

COMPANY: _____   PHONE: _____

TRUCK NO. _____   DRIVER: _____

TRAILER NO. _____   DRIVER: _____

DATE LEFT: _____   DATE RETURNED: _____

TRIP FROM: **TRIP TO:**

DATE	STATE	ODOMETER BEGINNING MILES	ODOMETER ENDING MILES	HIGHWAYS USED EACH STATE	STATE MILES

FUEL PURCHASE RECORD

DATE	NUMBER	GALLONS	COST	NAME	CITY	STATE	CASH/CHARGE CREDIT CARD

TRUCKER LOG

COMPANY: _____   PHONE: _____

TRUCK NO. _____   DRIVER: _____

TRAILER NO. _____   DRIVER: _____

DATE LEFT: _____   DATE RETURNED: _____

TRIP FROM: **TRIP TO:**

DATE	STATE	ODOMETER BEGINNING MILES	ODOMETER ENDING MILES	HIGHWAYS USED EACH STATE	STATE MILES

FUEL PURCHASE RECORD

DATE	NUMBER	GALLONS	COST	NAME	CITY	STATE	CASH/CHARGE CREDIT CARD

TRUCKER LOG

COMPANY: _____   PHONE: _____

TRUCK NO. _____   DRIVER: _____

TRAILER NO. _____   DRIVER: _____

DATE LEFT: _____   DATE RETURNED: _____

TRIP FROM: **TRIP TO:**

DATE	STATE	ODOMETER BEGINNING MILES	ODOMETER ENDING MILES	HIGHWAYS USED EACH STATE	STATE MILES

FUEL PURCHASE RECORD

DATE	NUMBER	GALLONS	COST	NAME	CITY	STATE	CASH/CHARGE CREDIT CARD

🚛 TRUCKER LOG

COMPANY: _____ PHONE: _____

TRUCK NO. _____ DRIVER: _____

TRAILER NO. _____ DRIVER: _____

DATE LEFT: _____ DATE RETURNED: _____

TRIP FROM: TRIP TO:

DATE	STATE	ODOMETER BEGINNING MILES	ODOMETER ENDING MILES	HIGHWAYS USED EACH STATE	STATE MILES

FUEL PURCHASE RECORD

DATE	NUMBER	GALLONS	COST	NAME	CITY	STATE	CASH/CHARGE CREDIT CARD

TRUCKER LOG

COMPANY: _____ PHONE: _____

TRUCK NO. _____ DRIVER: _____

TRAILER NO. _____ DRIVER: _____

DATE LEFT: _____ DATE RETURNED: _____

TRIP FROM: **TRIP TO:**

DATE	STATE	ODOMETER BEGINNING MILES	ODOMETER ENDING MILES	HIGHWAYS USED EACH STATE	STATE MILES

FUEL PURCHASE RECORD

DATE	NUMBER	GALLONS	COST	NAME	CITY	STATE	CASH/CHARGE CREDIT CARD

Made in the USA
Monee, IL
16 August 2022

11747314R00057